A Guided Philosophical Discussion

J. EDGAR THOMSON

A Guided Philosophical Discussion
by J. Edgar Thomson

© 2017 by J. Edgar Thomson. All rights reserved.

No part of this book or the concept of this book may be reproduced in any written, electronic, recording, or photocopying form without written permission of the author for reason of mass distribution. Concepts similar to historical or classical quandaries in the public domain may not be specifically duplicated without specific request. Permission shall be granted upon request.

Books may be purchased in quantity and/or special sales by contacting the printer, CreateSpace, an Amazon.com Company, at www.Amazon.com.

Published by:	Amazon.com, Inc., Seattle, WA
Written by:	P.O. Box 85, Long Beach, NY 11561
Cover Design by:	J. E. Thomson
Bowker ISBN:	978-0-9987733-2-2
CreateSpace ISBN:	978-1545230817
Subject:	1. Philosophy; 2. Discussion; 3. Questions
Edition:	First
Printed in:	the United States of America

My family who, for generations, allowed active discussion and debate to liven our table and sharpen our minds.

And with deep thanks to my father
Iron sharpens iron, and one man sharpens another

HOW TO USE THIS BOOK

This book is to be used as a simple guide for discussion. Room is left following each question to allow the reader to pen his own thought, or any clever realizations. Occasionally an additional question is included to prompt further thought. The proper location for this book is where one sits to enjoy their afternoon coffee and not at their bedside.

BE REMINDED of the following:
- ➢ The purpose of discussion is to encourage the open and active exchange of ideas
- ➢ Be curious in one another's thoughts and your minds will be opened to the active and fertile breeding of new ideas
- ➢ Either-Or reasoning leaves no room for an opposing view to have a shred of truth, Both-And reasoning allows for an infinite room of truth on either side
- ➢ Ask the question, "Why do you believe…?"

Does God exist?

J. Edgar Thomson

Who are you?

A GUIDED PHILOSOPHICAL DISCUSSION

What makes an individual human different from another?

J. Edgar Thomson

What is truth?

A GUIDED PHILOSOPHICAL DISCUSSION

What determines truth?

J. Edgar Thomson

What makes a person human?

A GUIDED PHILOSOPHICAL DISCUSSION

Can humans produce artificial intelligence? Will the created person have rights?

J. Edgar Thomson

What is intelligence?

A GUIDED PHILOSOPHICAL DISCUSSION

What is knowledge?

J. Edgar Thomson

Where are memories stored?

How do brain cells regenerate, but memories are not forgotten?

J. Edgar Thomson

Can one learn anything from ancient documents? How if the knowledge has been built upon?

A GUIDED PHILOSOPHICAL DISCUSSION

Can we experience anything without the senses?

J. Edgar Thomson

Is there any reason to fight if there is no chance of success?

A GUIDED PHILOSOPHICAL DISCUSSION

If one cannot influence their environment, does one die?

J. Edgar Thomson

Is euthanasia ever acceptable or moral?

A GUIDED PHILOSOPHICAL DISCUSSION

What is death?

J. Edgar Thomson

When one sleeps, does one die?

A GUIDED PHILOSOPHICAL DISCUSSION

Are you asleep?

J. Edgar Thomson

Are you dreaming?

A GUIDED PHILOSOPHICAL DISCUSSION

What is the difference between living and being alive?

J. Edgar Thomson

A hammer has the handle replaced, sometime later, the head of the hammer is replaced. Is it the same hammer?

A GUIDED PHILOSOPHICAL DISCUSSION

If your cells replace themselves every seven to ten years, are you the same person?

J. Edgar Thomson

How can the individual cells in a person be alive, but the individual person possess a separate life?

A GUIDED PHILOSOPHICAL DISCUSSION

Why are humans alive?

J. Edgar Thomson

What is the purpose of humanity?

A GUIDED PHILOSOPHICAL DISCUSSION

Does every individual life have a purpose?

J. Edgar Thomson

Will humans go extinct?

A GUIDED PHILOSOPHICAL DISCUSSION

Where does the soul reside?

J. Edgar Thomson

Where do thoughts come from?

A GUIDED PHILOSOPHICAL DISCUSSION

Are thoughts words or images? What are thoughts?

Are the brain, mind, and soul different? If so, how?

A GUIDED PHILOSOPHICAL DISCUSSION

What is a soul?

J. Edgar Thomson

How does mind and body interact?

A GUIDED PHILOSOPHICAL DISCUSSION

Is a person a mind or a body?

J. Edgar Thomson

What is consciousness?

A GUIDED PHILOSOPHICAL DISCUSSION

Why can humans reason?

Who decides what is moral? Why?

How does morality change based on factors outside of our control?

J. Edgar Thomson

Does right and wrong change based on the opinion of society?

A GUIDED PHILOSOPHICAL DISCUSSION

Are women or men or valuable to society?

J. Edgar Thomson

Does right and wrong very between societies?

A GUIDED PHILOSOPHICAL DISCUSSION

How do people determine if their actions are right?

J. Edgar Thomson

Are people in control of their lives?

A GUIDED PHILOSOPHICAL DISCUSSION

Is doing something wrong acceptable if no one knows?

J. Edgar Thomson

Is any action performed free of consequence?

A GUIDED PHILOSOPHICAL DISCUSSION

What role does honor play in society? How is it expressed?

J. Edgar Thomson

Is there a time when honor is not acceptable?

A GUIDED PHILOSOPHICAL DISCUSSION

Why is something beautiful rather than ugly?

J. Edgar Thomson

Why is something ugly rather than beautiful?

A GUIDED PHILOSOPHICAL DISCUSSION

If one cannot buy happiness, can one without money be truly happy?

J. Edgar Thomson

Would you live in a computer simulation if it would increase your happiness?

A GUIDED PHILOSOPHICAL DISCUSSION

Is it better to be an unhappy human or a happy animal?

J. Edgar Thomson

How do you know your perceptions are real?

A GUIDED PHILOSOPHICAL DISCUSSION

Why does money have value?

J. Edgar Thomson

Is monetary inflation ever acceptable?

A GUIDED PHILOSOPHICAL DISCUSSION

How much control do you have over your own life?

J. Edgar Thomson

What is life?

A GUIDED PHILOSOPHICAL DISCUSSION

Is there a meaning to life?

J. Edgar Thomson

What is freedom?

A GUIDED PHILOSOPHICAL DISCUSSION

Can persons freely sell themselves into slavery?

J. Edgar Thomson

Is slavery ever acceptable?

Can a person reasonable perform an action if it interferes with another's freedom?

J. Edgar Thomson

How does freedom differ from liberty?

A GUIDED PHILOSOPHICAL DISCUSSION

If one could live forever, would they?

J. Edgar Thomson

How is life measured?

A GUIDED PHILOSOPHICAL DISCUSSION

Are animals more valuable than humans?

Why is animal abuse offensive?

A GUIDED PHILOSOPHICAL DISCUSSION

Is abortion ever acceptable?

J. Edgar Thomson

Is suicide ever acceptable?

A GUIDED PHILOSOPHICAL DISCUSSION

Is it acceptable to kill ten people to save a thousand people?

J. Edgar Thomson

If by pressing a button one would receive more wealth than one could spend, but kill ten people, would you press the button? If so, how often?

A GUIDED PHILOSOPHICAL DISCUSSION

How much money would you sell a day of your life for? How much would you sell a year for? If not, why do humans work?

J. Edgar Thomson

Do different lives have different value?

A GUIDED PHILOSOPHICAL DISCUSSION

If one can travel back in time, can one kill oneself?

J. Edgar Thomson

What is after life?

A GUIDED PHILOSOPHICAL DISCUSSION

Why do we respect the Dead?

J. Edgar Thomson

What is the purpose of laws?

A GUIDED PHILOSOPHICAL DISCUSSION

Is loving easier than being loved?

J. Edgar Thomson

Is trust more important than love?

A GUIDED PHILOSOPHICAL DISCUSSION

Is forgiving different than forgetting? Do the two interact?

J. Edgar Thomson

Is morality relative?

A GUIDED PHILOSOPHICAL DISCUSSION

Is humanity becoming more moral with time?

Is patriotism irrational?

Is race a social structure or a biological reality? What is the case for multi-racial persons?

J. Edgar Thomson

Is a person's perception of reality the same as one's experience of reality?

A GUIDED PHILOSOPHICAL DISCUSSION

Were you the same person ten years ago?

J. Edgar Thomson

What is freewill?

If an all-knowing god exists, is there freewill? If freewill does not exist, why do we punish people for committing crimes?

Should high-risk individuals be jailed prior to committing a crime?

A GUIDED PHILOSOPHICAL DISCUSSION

At what point is a crime committed?

Do animals have freewill? Do animals have emotions?

A GUIDED PHILOSOPHICAL DISCUSSION

What rights do animals have?

J. Edgar Thomson

Do plants have freewill? Do plants have emotions?

A GUIDED PHILOSOPHICAL DISCUSSION

What rights do plants have?

What is occurring beyond observable space? Why does it occur? Why are their storms on Jupiter?

Why do humans live for reasons other than self-gratification?

Does something exist that cannot be perceived or measured?

A GUIDED PHILOSOPHICAL DISCUSSION

Does evil exist? As something separate from individual choice?

J. Edgar Thomson

Are love and passion different?

A GUIDED PHILOSOPHICAL DISCUSSION

Does every event have a cause?

J. Edgar Thomson

Should people receive more votes if they pay more taxes?

A GUIDED PHILOSOPHICAL DISCUSSION

Does art and music affect emotions? Why?

J. Edgar Thomson

Are emotions practical?

A GUIDED PHILOSOPHICAL DISCUSSION

Are emotions constructive to civilization?

J. Edgar Thomson

If one refuses to save another, are they responsible for the death?

A GUIDED PHILOSOPHICAL DISCUSSION

If pain exists only in the brain, does it exist at all?

J. Edgar Thomson

Why is there something and not nothing?

A GUIDED PHILOSOPHICAL DISCUSSION

From the point of view of human nature: Is there an alternative to Capitalism (free market)?

J. Edgar Thomson

Can a person experience the same thing twice?

A GUIDED PHILOSOPHICAL DISCUSSION

What is the most important moment in a human's life?

J. Edgar Thomson

Why can we perceive the world around us?

A GUIDED PHILOSOPHICAL DISCUSSION

Can a blind man, made sighted, perceive things by sight after learning them by touch?

J. Edgar Thomson

Is color possessed by a particular object, or does it exist in the mind?

A GUIDED PHILOSOPHICAL DISCUSSION

Do people see the same colors equally, or are colors individually perceived?

J. Edgar Thomson

Can parts of a whole be removed without changing the whole?

ABOUT THE AUTHOR

The author respectfully wishes you the best in your discussions and debate, may the completion of this book bless you with the energy to revisit your previous answers and see if your thoughts have changed. The author also invites you to send him your thoughts, feelings, and opinions. He is always up for a fight.

www.ingramcontent.com/pod-product-compliance
Lightning Source LLC
Chambersburg PA
CBHW020920180526
45163CB00007B/2823